MANEATERS and Blood Suckers

KIRSTY MURRAY

A LITTLE ARK BOOK

ALLEN & UNWIN

Copyright © text Kirsty Murray 1998
© illustrations Craig Smith 1998

All rights reserved. No part of this book may be transmitted in any form or by any means, electronic or mechanical, including photocopying, recording or by any information storage and retrieval system, without prior permission in writing from the publisher.

First published in 1998
A Little Ark Book
Allen & Unwin Pty Ltd
9 Atchison Street
St Leonards NSW 2065, Australia
Phone: (61 2) 9901 4088
Fax: (61 2) 9906 2218
E-mail: frontdesk@allen-unwin.com.au
URL: http://www.allen-unwin.com.au

10 9 8 7 6 5 4 3 2 1

National Library of Australia
cataloguing-in-publication data:

Murray, Kirsty.
　Man-eaters and blood-suckers.

　Includes index.
　Includes glossary.
　ISBN 1 86448 630 9.

　1. Dangerous animals. 2. Human–animal relationships. I. Title. (Series: True stories (St. Leonards, N.S.W.)).

591.65

Photo credits
Cover photo & the photo of grizzly bear, International Photo Library
Bengal tiger, great white shark and leeches, ANT Photo Library
Crocodile hunt, Coo-ee Historical Picture Library
Alf Collins photo used courtesy of the Royal Humane Society of Australasia

Thanks to Catherine O'Rourke for help with photo research

Typeset by Midland Typesetters, Maryborough, Victoria
Printed in Australia by McPherson's Printing Group, Maryborough, Victoria

Contents

Introduction *Facing up to fear*   **8**

1 **Big Cats**   **9**
Big cat breakfast • Who's eating who? • Making a meal of it • The Sundarban man-eaters • Say your prayers

2 **Wolves**   **18**
Wolf plague • Who is the Big Bad Wolf? • The Beasts of Gévudaun • Where wolves? • Wolf women • Wolf children

3 **Bears**   **27**
If you go into the woods today... • Bears' picnic • The bear facts • No photographs – please! • Polar giants • The last hunt

4 **Ferocious Ferals & Dangerous Domestics**   **37**
Pigging out • Having a boaring time • Diabolical dogs • Best friend or worst enemy? • Dancing with dingoes • Rats on the rampage • Cats meow • Too much monkey business • The last word on mammals that maul

5 **Reptiles**   **47**
Home for the horrordays • Rampage against reptiles • A Nile crocodile's evening snack • Crocodile rock 'n' roll • Sussing out serpents • People and pythons • When a dream becomes a nightmare

6 Creatures of Deep (and not so deep) Waters   58
The big bite • Shaping up to sharks • A hero's death • Shark techniques and other tidbits • The tattoo mystery • The long arm of death • Picking on piranha • The problem with getting wet

7 Parasites   68
The most successful man-eaters of all • Getting around roundworm • A pick and mix of parasites • The tapeworm's tale • Hookworms get their hooks in • Wormy waters

8 Wings, Stings and Slimy Things   76
Winged nightmares • The flying dustmen • Sky burial • The thing in the sting • Not-so-nice lice • Tickles for Tim • Leeches lick it up

9 Cannibals   85
The makings of a murderous meal • The last voyage of the Mignonette *• Hunting up cannibals • Staying alive*

Glossary	**92**
Further Reading	**94**
Index	**96**

Acknowledgements

Thanks to Ken for inspiration and encouragement, and to all the gang – Ruby, Billy, Elwyn, Isobel, Romanie and Theo – for putting up with months of unappetising dinnertime conversation.

Thanks also to John Murray for interesting anecdotal information and to Patrick Wolfe for his illuminating commentary on cannibalism. Special thanks to Sarah Brenan for her nerves of steel in reading through the many drafts of these stories and helping me render them more 'palatable'.

Introduction

FACING UP TO FEAR

As a kid, I was really scared of animals – especially ones that bit. One of my earliest nightmares was of my big sister transforming into Disney's big bad wolf and chasing me around the house before gobbling me up in the front yard.

Fear is a very sensible emotion – it can save your life – but when your fears get out of perspective, they can make you miserable. So if some of the stories in this book begin to get to you, remember they are not typical. Human beings are not the natural prey of any animal except parasites and insects. You are more likely to be eaten by your next door neighbour than by the big bad wolf. In fact, by the time I finished researching this book, I'd realised that animals actually have a lot more to fear from us than we do from them. Human beings are the ultimate predators – we eat a wider range of creatures than any other carnivore on the planet.

Learning to respect and understand the things you are afraid of can really help you get a handle on your fears and make you a whole lot braver – well, that's what writing this book has done for me!

1 Big Cats

BIG CAT BREAKFAST

Nepal, April 1992
Prem, Uday and Moti were herding cattle along a trail that led into the forest. Slightly ahead of them, 12-year-old Dol Ram Mahato was alone with his cattle – at least that was what the younger boys thought, until they saw his body lying a few metres from the path in a pool of blood. The Chepte tigress had claimed another victim.

The three boys turned and bolted back towards the village. The narrow path was surrounded by dense undergrowth and they were forced to run in single file, Moti bringing up the rear. Moti cried out only once, a long high scream, as the tigress leapt on him and sank its 7-cm teeth through the top of his skull. The two other boys turned to see their friend in the jaws of the tiger, being dragged back along the trail to where Dol Mahato's body lay.

It took Prem and Uday ten minutes to get to the village, where they raised the alarm. A group of villagers went to rescue the bodies of Dol and Moti. It was too late to save their lives, but at least they would get a decent funeral.

A month later the Chepte tigress was caught,

MAN-EATERS AND BLOOD-SUCKERS

brought to ground by two darts that were meant to knock her out so she could be transported to a zoo. She died of complications from the dart wounds.

WHO'S EATING WHO?

More people in the world are killed by domestic dogs than eaten by big cats (see Chapter 4), but when you think of four-legged man-eaters it's tigers, not poodles, that spring to mind. Nature designed the big cats as hunters – amongst the most efficient killing machines on the planet. They're bigger, faster and in some ways even smarter than we are.

> More people ... are killed by domestic dogs than eaten by big cats

You'd think we'd make a tasty snack – no fur, no nasty prickly bits to peel off like porcupines – and we're not half as hard to catch as an antelope. But the fact is, people figure at the bottom of the list of preferred dinners for all the big cats. We can't taste half as good as we look. For the most part, they avoid us as if we were a dose of very nasty medicine. Only a few species occasionally put human on the menu – lions, leopards, pumas and tigers. Cheetahs, despite their size and speed, are not the least bit interested in us – they probably think we're so slow we're not worth bothering to chase.

In fact, humans are much more dangerous to cats than the other way round. There are 36 species of wild cats around the world, and nearly all of them are considered threatened species. Many of them are at risk of becoming extinct. Even though they are

Big Cats

protected by law in most countries, they face the threat of poachers and the destruction of their forest homes. In 1900 there were probably 100 000 tigers spread across Asia – now, there are fewer than 8000 world-wide. Of eight species that existed 50 years ago, three – the Caspian, Java and Bali – are now extinct. In the course of the past few years there's been a surge of illegal trading in tiger body parts for use in traditional Chinese medicine. Every year, hundreds of tigers are poached, so their bones, skin and meat can be sold on the black market. The bones are used to make a huge range of products from tiger-bone wine to lotions and potions for every ailment you can think up. In Taiwan, tiger stew features on the menu in illicit gourmet restaurants that specialise in serving up endangered species to their patrons. So who's eating who?

MAN-EATERS AND BLOOD-SUCKERS

Reasons for big cat attacks
- You've come between them and their cubs.
- Bad timing – you've disturbed them with a fresh kill.
- They are wounded and can no longer hunt their natural prey, like Shere Khan of Kipling's *Jungle Book*. This is the most common reason for turning to man-eating.
- Their habitat (where they live) has been completely overrun with people.
- They're cornered.
- They're starving.
- They're bored.

Making a meal of it

In 1898 a pair of male lions killed and ate approximately 140 people who were struggling to construct a railroad bridge across the Tsavo River in Kenya. (No one knows the exact number killed – estimates range from 40 to 200.) The project leader, Colonel J. H. Patterson, couldn't deter these animals, nicknamed 'The Ghost' and 'The Darkness' by local people. Even the experts – the Masai lion hunters – gave up!

Eventually, Colonel Patterson managed to shoot them both, and work on the bridge continued. The lions were stuffed and mounted, and are on display at Chicago's Field Museum of Natural History.

A lion can eat 34 kg of meat in a single sitting, so an adult human could make a tidy meal for two. Yet man-eating lions are pretty uncommon – the Tsavo pair may have been driven to man-hunting by famine.

Big Cats

> **DID YOU KNOW?**
> The African male lion grows to 3 m long (including the tail) and 1 m high, and weighs up to 230 kg. He can span nearly 9 m at one bound, jump over a barrier almost 2 m tall, and dash a short distance at more than 50 km per hour.

Unlike us, big cats never hunt for sport. They eat what they kill, though occasionally, like a house cat, they might play with it first. All big cats start on the intestines and work their way up and down the length of the body, leaving little to waste. Jim Corbett, one of the world's best-known hunters of man-eaters, often reported finding only the head or feet of the big cat's victim. Corbett stalked some of India's most infamous man-eaters. He grew to love and respect his prey, and eventually set up a reserve to save the tiger. He understood whose lives were really under threat.

> All big cats start on the intestines

THE SUNDARBAN MAN-EATERS

Sanyasi Mandal, a young Indian fisherman, awoke one night to feel the hot breath of a tiger on his face. In the flickering light of a kerosene lamp, the tiger's eyes glowed. Suddenly it seized the 40-year-old man who lay beside Sanyasi and leapt over the side of the boat. It swam to shore as silently as it had arrived, with its unlucky victim in its jaws.

Sanyasi and his three companions were fishing in the swampy Sundarban region of West Bengal, where

MAN-EATERS AND BLOOD-SUCKERS

three great rivers meet – the Ganges, Brahmaputra and Megna. These rivers have created a delta of 80 000 square kilometres and the largest mangrove forests in the world. No one lives there, but every year thousands of people enter the winding tidal creeks and thick tangle of mangrove forest to fish, cut timber and collect honey from nests of wild bees. And every year up to 50 of them are eaten by the tigers of the Sundarban.

Only a dozen of the Sundarban's 250 tigers are man-eaters. They are the exception to the rule of big cats as innocent killers – there's a theory that the salty water of the swamp makes them particularly aggressive.

Wearing rubber masks has become compulsory for mangrove visitors. Tigers track by stealth and leap at their prey from behind, sinking their teeth into the neck or skull of their victim. A face mask placed on the back of your head may just confuse a tiger

Big Cats

enough to keep him at bay. Even so, forest workers using these masks have been followed for up to eight hours by hungry tigers; and in the last few years, some of the animals seem to have got wise to the trick and have worked out the difference between masks and the real thing.

Another way of deterring the tigers of the Sundarban is to put up clay dummies of people – honey collectors and wood cutters – in the forest and link them up to 230 volts of electricity. This is meant to teach the tigers that messing with humans is a shocking business, but the message isn't getting through. Still, the tigers have developed a healthy respect for firearms, and never attack the patrol boats of the forest guards.

> **DID YOU KNOW?**
> Siberian tigers are the biggest of all tiger species. They can tilt the scales at 225 kg! Tigers can be 2.8 m long – add an extra 1.2 m for the tail. They are powerful swimmers and can easily cross rivers as wide as 6 km with their lunch in their jaws.

SAY YOUR PRAYERS

The Convent of San Francisco lay on the banks of the Rio Grande in California, 27 km from Sante Fe. On a warm spring morning in 1825, Brother Ramon

MAN-EATERS AND BLOOD-SUCKERS

was busy tidying up the chapel after Mass. He breathed a little sigh of satisfaction when all was sorted and opened the door to the sacristy to put away the wine.

> He might have thought it was the devil himself

He might have thought it was the devil himself leaping at him from the small dark room, though he probably didn't have enough time to think anything. He let out his last scream and alerted the whole of the convent to the presence of a jaguar.

In the course of the morning two more brothers met the same fate as Brother Ramon.

As the monks sat terrified in the chapel, Brother Francis had a brainwave. Opening onto a small enclosed garden at the back of the convent was a second door to the sacristy. Brother Francis decided that if he could get this door open, the beast might leave them in peace. Unfortunately for the monk, the jaguar had exactly the same idea and the two crossed paths in the convent garden.

The jaguar claimed his fourth victim before leaping the wall of the garden and disappearing into the forest.

There are no longer any jaguars left in California (except the type with four wheels) to prey upon monks or anyone else. They've been exterminated in the United States, most of Mexico and all but a small corner of Argentina. Like most of their relatives, they're an endangered species.

Big Cats

How to survive a big cat attack

If you're unlucky enough to be nabbed from behind, there's probably not much you can do about it but relax and hope someone will come to your rescue. Once he's got you in his jaws, it probably won't hurt as much as you would imagine. Dr David Livingstone, a British explorer attacked by a wounded lion, reported feeling little pain as the lion munched on his shoulder. There's a theory around that the deep purring that big cats make acts as a natural sedative on their victims, causing them to struggle less and making it easier to finish them off.

The rule of thumb with the big cats is to **watch your back**. **Keep upright** – people cutting grass are often number 1 targets because they have their backs turned and are bent over. Kids are also preferred snacking material – whatever respect a big cat may develop for a grown-up, it will always treat a child as legitimate prey or a plaything. So **look tall**.

And keep your cool. Remember, unless he's really desperate, he's not interested in eating you. **Don't stare** – he might take it as a challenge – but do keep your eye on him. Sudden movements aren't a good idea. **Move slowly** but when things are looking tense, stand still and hold your ground – don't turn your back and retreat.

Whatever you do, don't lie down and play possum. He'll probably think the hard part is over and take you off to pop in some convenient spot for later snacking.

2 Wolves

WOLF PLAGUE

It was a quiet August evening in 1996, and ten-year-old Sita Devi and her mother were walking through a grassy clearing at the edge of their village. Sita was searching for a place to use as a toilet. Ahead of her was her four-year-old brother, Anand – looking towards a stand of corn and elephant grass.

Sita heard her mother scream.

A huge wolf appeared from nowhere, seized the small boy by the neck and disappeared with him into the corn, heading towards a nearby riverbank.

Three days later, a police search party found the remains of Anand's head a kilometre away.

In 1996, over 30 children were reported carried off and killed by wolves in the state of Uttar Pradesh. Ram Lakhan Singh, an animal conservationist, was chosen to lead the effort to kill the man-eating wolves. He believed that a single wolf pack was responsible and that hunger was the most likely cause of the attacks. Uttar Pradesh is one of the most densely populated states in India, and most of its forests have been destroyed to make way for farms. The small band of desperate wolves that remained was forced to live on the

Wolves

edge of villages as old hunting grounds disappeared.

Like big cats, wolves prefer to give people a wide berth. They know who is the most ferocious hunter, and no healthy wolf really wants to mess with the top dog. They are much more interested in hunting deer than snatching babies, but when the animals they normally hunt disappear because of the spread of farms and villages, they get pretty desperate in their search for food.

> **Like big cats, wolves prefer to give people a wide berth.**

Over 100 years ago a similar 'wolf plague' occurred in India, and the government offered villagers a bounty for wolf pelts. More than 2600 wolves were slaughtered in less than nine months. Conservationists are worried that if bounties are reintroduced, the Indian wolf will be wiped out.

WHO IS THE BIG BAD WOLF?

Some conservationists believe wolf hybrids (half dog, half wolf) or wild dogs are more likely to be child-killers than pure wolves. Wolf hybrids are often bigger than pure wolves, and they are also less timid.

Whatever killed Anand Devi – whether it was a werewolf, wolf, wild dog, or wolf hybrid – his death touches on one of the deepest conflicts between people and beast. Around the world, children are told tales of the 'Big Bad Wolf'. He stalks our nightmares – the villain of 'Little Red Riding Hood' and hundreds of other folk tales and horror stories.

MAN-EATERS AND BLOOD-SUCKERS

People have imagined he possesses supernatural powers – wolves are often portrayed as a symbol of death and destruction and the companions of Dracula. But the wolf has had a lot of bad press.

It's true that wolves do not make good neighbours or pets. They are wild animals and they hunt other animals for food. Yet wolves and people actually have a lot in common. Some

> wolves and people actually have a lot in common

anthropologists believe that we originally learnt how to hunt by observing wolves. Like us, they are highly social animals and live in family groups, called

DID YOU KNOW?
Every dog in the world is descended from wolves that were tamed in the Middle East about 12 000 years ago. Thousands of generations of careful breeding has changed the built-in behaviour of dogs. Although they may seem to have a lot in common with dogs, wolves are different. They are predators that belong in the wilderness.

Wolves

> **DID YOU KNOW?**
> The ancient Greeks believed the god Zeus often took on the form of a wolf when he visited Earth. In Athens, any man who killed a wolf had to pay for a wolf funeral.

packs. A pack is really an extended family with between two and 30 members in it. Wolves are gentle and affectionate parents. Perhaps that's why there are so many legends of human children being raised by wolves, such as Mowgli of *The Jungle Book*. No other animal has such a complicated reputation, or such a troubled relationship with humans.

THE BEASTS OF GÉVUDAUN

Probably the most famous man-eating wolves of history were a pair of grey timber wolves called the Beasts of Gévudaun. They roamed the Cévennes Mountains of south-central France, preying on livestock. In June 1764 they took their first human victim, and in the course of the next three years killed at least 64 people and possibly as many as 100. Most of their victims were small children.

People became so terrified that King Louis XV sent a series of small armies to bring the wolves to ground. Forty-three thousand soldiers and 2300 dogs spent two months trying to catch the desperate pair, with no success.

Finally, on 21 September 1766, an old peasant

MAN-EATERS AND BLOOD-SUCKERS

hunter called Antoine de Bauterne shot the male wolf. By European standards it was enormous. It weighed 59 kg, stood 81 cm at the shoulder and measured nearly 174 cm from nose to tail. It took another nine months to track down and kill the female.

The Gévudaun wolves were hard to catch, but their relatives were not so elusive. Over the next 30 years, more than 2000 wolves were killed in the Cévennes. There are no longer any wild wolves in France, or for that matter most of Europe.

In 1991, the movie-star-turned-environmentalist Brigitte Bardot rescued 40 Mongolian wolves that an irresponsible wildlife dealer was trying to sell on the

Wolves

European black market. Many of the wolves died, but Ms Bardot established a Wolf Park at Gévudaun in memory of the Beasts.

On the way to Granny's house ...
Little Red Riding Hood could have saved herself a lot of grief if she'd known a bit more about the Big Bad Wolf. Here are a few tips that might come in handy if you ever cross a wolf's path.

- Make it clear that you're not worth messing with. Stand tall – a wolf won't tangle with someone big, brawny and upright.
- Stay calm – sudden, jerky movements are definitely a bad idea.
- Don't turn your back, but don't try and stare the animal down either.
- Shouting in a low, deep voice could help, but screaming is out.
- If you've got something to fling, definitely throw it. If you can arm yourself with rocks and sticks without having to crouch down to get your ammunition, the wolf will get the message.
- Climbing a convenient tree is a good last-ditch way to save yourself, though you could have a bit of a wait before the wolf clears off.

WHERE WOLVES?

There have been very few attacks on humans in Europe or the Americas in the last 150 years. This is partly because there are no longer many wolves

MAN-EATERS AND BLOOD-SUCKERS

living anywhere near humans. Except on the Russian steppes and in isolated pockets of Scandinavia, there are few wolves left in Europe. There are probably fewer than 2500 wild wolves left in the whole of the United States, most of them in Minnesota.

Until about 50 years ago, many people believed the world would be a better place without any wolves in it. Wolves were slaughtered in huge numbers – trapped, poisoned and ruthlessly shot. In huge areas of North America, they were completely wiped out.

WOLF WOMEN

Patricia Wyman loved wolves. Her life ambition was to work with them. She studied wildlife biology at the University of Guelph in Ontario, Canada. Fresh out of college, she landed her dream job. Five timber wolves were kept in a six-hectare wooded enclosure at the Wolf Centre in the Haliburton Forest Wildlife Reserve. Tricia was hired to act as their caretaker and to run the education program at the centre.

On Monday 14 April 1996, Tricia took up her new position. Four days later, on Thursday afternoon, she entered the wolf enclosure for the third and last time. Her body was discovered later in the afternoon by two other employees. The wolves had torn off her clothing, and she was covered in bite wounds from which she had bled to death. When police officers entered the compound to remove her body, the wolves began circling them until warning shots were fired. The next day all five wolves were killed.

Wolves

Patricia Wyman's tragic death is difficult to explain. The wolf pack was effectively wild and usually kept its distance from any people who entered the enclosure. Wolf experts suggest Tricia may have tripped over a fallen branch, and once down was attacked by the wolves, who were guarding their territory from strangers.

North American Indians have many legends about wolf-women; women who go to live amongst wolves and learn magic and wisdom from them. The Mexicans told tales of *La Loba* – an old wolf-woman who gathered up bones, especially those of wolves, and sang them back to life. Legends are slippery things, but there is often a kernel of truth in them. Wolf Parks and Centres are probably the best way we can unravel some of the mystery of the way of the wolf.

WOLF CHILDREN

Judging from the stories above, the only real wolf children are in the bellies of wolves. Mowgli of *The Jungle Book*, the most famous human wolf-cub in fiction, is probably just that – a fiction. But whatever the facts, the legends abound, evidence of the endless fascination people have with wolves.

In every country where there have been wolves there are legends about their human cubs. Probably the most famous 'wolf child' was Victor, the Wild Boy of Aveyron. When he was eleven years old, he was found running naked and wild in the forests of Aveyron. He was captured on 25 July 1799 and taken to Paris, where he was lovingly cared for by

MAN-EATERS AND BLOOD-SUCKERS

> **DID YOU KNOW?**
> Romulus and Remus, the legendary twin founders of ancient Rome, were meant to have been raised by a she-wolf. Tu Kueh, the founder of Turkey, was supposedly suckled by a wolf. So was Siegfried, the Teutonic warrior, and Zoroaster, a sixth-century religious reformer.

Jean Itard, the chief medical officer of the Institute for the Deaf and Dumb. Jean Itard spent the next six years trying to teach Victor to speak – without much success. Some good came of it, though – a lot of the work Dr Itard did with Victor led to new ways of helping intellectually disabled children. Most researchers now believe that Victor had not been raised by wolves, but had been abandoned by his parents because of his disabilities.

The patron saint of wolves is Saint Ailbe, a sixth-century Irish bishop. The son of a slave girl, he was abandoned in the wilderness as a baby and raised by a she-wolf. Years later, when he was Bishop of Emily, a grey wolf who was being pursued by hunters ran into his home and laid its mighty head on his lap.

'I will protect thee, old mother,' he said as he drew his cloak around her. Out of love and respect for the old wolf, Saint Ailbe housed and fed her for the rest of her days.

Whatever the truth may be behind the legends, they help us to understand what our relationship with wolves should be.

3 Bears

IF YOU GO INTO THE WOODS TODAY ...

In August 1984, Barbara Moore had an unforgettable surprise. She and her husband, Bob, were hiking through the forest, when they came across the body of a dead bison. Barb was a ranger in a national park and she knew to keep her distance from dead animals in grizzly bear country – she didn't want to wind up coming between a bear and its breakfast. The couple stayed at the edge of the clearing, observing the area through binoculars.

'Barb, Barb, look out!' cried Bob, a split second before Barbara turned to see a mother bear charging her with two little cubs in pursuit. Hoping the bear was only bluffing, Barbara held her ground; but the bear grabbed her by the left leg and flung her face downwards to the ground. As the bear's teeth and claws slashed her back and thighs, Barb quickly realised her best chance of surviving was to play dead. Immediately, the bear lost interest and went for Bob, biting him on the knee before charging back to attack Barbara again.

Barb lay face down and motionless where the bear had left her. The bear reached down and rolled her over. Barb used the momentum of the push to keep on rolling until she was lying face down. Three times, the bear attempted to roll Barb over so she

MAN-EATERS AND BLOOD-SUCKERS

would be face upwards. Each time, despite her injuries – severe head wounds, a lacerated thigh and multiple puncture and claw marks – Barb kept on rolling until she was face down, hardly daring to draw breath as she did so. Finally, after the failed third attempt, the bear gave up and wandered off with her cubs.

Barbara's know-how and ability to stay calm in the face of incredible danger saved her life. Bear attacks are usually sudden close encounters and are over quickly unless the bear is injured.

Bears are maulers who are extremely territorial and defensive of both their young and their dinner. Barbara's encounter was a recipe for disaster – she had accidentally come between a bear and its kill. Worst of all, it was a she-bear who would be doubly defensive because of her babies. Although mother bears with cubs make up only 20 per cent of any bear population, they account for more than 80 per cent of bear attacks.

> Barbara's encounter was a recipe for disaster – she had accidentally come between a bear and its kill.

BEARS' PICNIC

Bears are omnivorous – they eat anything and everything they can lay their paws on. Considering this, it's surprising they don't try and make a meal of us when they have the chance. Humans would be easy to catch and kill, easy to eat, and there are a lot of us around – yet bears express little or no interest in eating us.

One rare exception to that rule is the case of the unlucky Assiniboin Tribe of Stonie Indians. In 1860 the tribe was struck down by a smallpox epidemic while camping at Yarrow Creek, Alberta. Grizzlies, being scavengers, began feeding on the bodies of the dead. When the Indians moved on to their next camp, the bears began preying on the survivors, having acquired a taste for human flesh. For years after, the Stonie Indians avoided Yarrow Creek for fear of bumping into one of the grizzlies that had learned to prey on humans.

> **DID YOU KNOW?**
> During late summer, bears have been known to consume up to 20 000 calories per day. That's the equivalent of 40 hamburgers finished off with 40 icecream sundaes in a single day!

THE BEAR FACTS

Bears are found on every continent except Australia. (The koala is not a true bear.) There are eight bear species and each one is endangered. The bears that

MAN-EATERS AND BLOOD-SUCKERS

> **Staying alive in bear country**
> The best way of avoiding bear conflict is to avoid bears.
> - Never go near potential feeding areas such as berry patches or dead animals.
> - Don't carry food or heavily scented soaps. Bears are very attracted to interesting smells.
> - Be noisy. Sing, talk and shout as you walk. If the bear knows you're coming, it will avoid you.
> - Leave your dog at home. A dog often infuriates a bear. It could come running back to you with a bear in pursuit!
> - Occasionally, a bear attack can be a case of mistaken identity. Try not to look, sound or smell like a prey species – wearing a fur coat could be a very bad idea.

most often have conflict with humans are the American brown bear, the American black bear, the Asian black bear, and the polar bear.

The American brown bear is responsible for the highest number of attacks. Barbara Moore was mauled by a grizzly, which is one variety of brown bear. Brown bears are famous for being both bad-tempered and unpredictable.

> **Brown bears are famous for being both bad-tempered and unpredictable.**

When brown bears attack they are defending their territory, not looking for dinner. If you come across a brown bear, talk to it and let it know you're a human – not another bear or a prey species. Stand tall, and let it see you for what you

Bears

are. Don't stare or make sudden movements that may threaten it.

If the bear is standing on its hind legs, it's just checking you out. When it gets down on all fours, it's time to take it seriously – bears about to attack will lay their ears back and charge on all fours; but even if it charges, there's a good chance it's bluffing. If it isn't a bluff, playing possum is the best tactic. Lie flat on your stomach, cover your head and neck with your hands and don't move until you are 100 per cent sure that the bear has left the area. The second attack is often worse than the first. If the bear has moved on, it's your cue to do likewise – in the opposite direction.

Playing possum with a brown bear is useful; but playing possum with a black bear is a big mistake. Black bears eat carrion (dead things). Out of curiosity, a black bear might just rip you open to check out what's inside. Black bears can bite through a tree trunk thicker than a man's arm, but in

MAN-EATERS AND BLOOD-SUCKERS

general they are not aggressive. Most black bear attacks are simply attempts to get food – if you've got it, hand it over.

Black bears are slightly smaller, and nowhere near as territorial or as frightened for their young as their cousins the brown bears. They can easily be scared off by aggressive behaviour. Shouting, yelling and throwing rocks or sticks will convince any black bear you're more trouble than you're worth. Just be sure you have the right bear, as black and brown bears can look suspiciously alike. Watch out for the claws – brown bears have 7-cm claws that are highly visible and they have larger heads than black bears.

Never try to outrun any type of bear. They can run as fast as horses both uphill and down – at speeds up to 65 km per hour! Climbing trees to escape a bear attack is not a great idea either. It's unlikely you'll get to the tree before the bear.

> **They can run as fast as horses both uphill and down**

Whatever type of bear you're up against, the cardinal rule of survival is **STAY CALM**. Staying calm is the best way of surviving any attack.

DID YOU KNOW?
People have occasionally survived bear attacks by scaling trees, but getting high enough is the first problem and staying up there can sometimes prove the second. At least two people in North America have been killed falling from trees during an attack. The bears showed no interest in them once they hit the ground.

Bears

NO PHOTOGRAPHS— PLEASE!

Bears are shy, solitary creatures. They don't like being approached. Photographers tend to receive the severest injuries of people who are mauled. Perhaps bears are camera-shy or, more likely, photographers can be too pushy.

In July 1994 an Australian woman holidaying in Anchorage, Alaska decided to get a closer look at Binky and Nuka, a pair of polar bears at the Alaska Zoo. Two sets of safety rails separated the public from the bears. In pursuit of the perfect picture, the woman climbed over the first set of rails. From Binky's point of view, this was too close for comfort and he lashed out at her, forcing his head through the bars and grabbing her leg in his jaws. The woman escaped with a broken leg. Locals made fun of the woman's misfortune and criticised her for climbing the safety rails. Her mistakes were being ill-informed, over-confident and unlucky.

Six weeks later, two drunken local teenagers climbed into Binky and Nuka's enclosure at night. They were planning a quick dip in Binky's pool. Binky showed his outrage by mauling the leg of one of the teenagers while he was undressing at the front of the enclosure. The teenager was lucky enough to be dragged to safety by his friend. Perhaps the kids thought it was only tourists that tickled Binky's tastebuds. A lack of respect for the power of predators can land you in serious trouble.

Perhaps the most appalling story of people not respecting the power of bears is of parents who rubbed jam on their 21-month-old baby's face and

MAN-EATERS AND BLOOD-SUCKERS

held him out the car window. They hoped to get the perfect 'Disneyland' shot of their baby being licked clean by a bear. You can guess the tragic result.

> **DID YOU KNOW?**
> In Churchill, a small town on the west coast of Canada's Hudson Bay 60 km from a large polar bear denning area, parents and armed motor patrols accompany kids on their Hallowe'en trick-or-treat visits.

POLAR GIANTS

Between 1965 and 1985, 20 people were mauled by polar bears in Canada – and six of these attacks proved fatal. Unlike other varieties of bears, polar bears are solid carnivores. Although they may scavenge in garbage dumps for odd titbits, these bears like meat and plenty of it – they are the largest meat-eating hunters on the planet. The biggest one ever recorded tipped the scales at 1002 kg – that's one tonne – and measured 3.5 m in length.

> **the largest meat-eating hunters on the planet**

Like grizzlies, polar bears are fierce. If you're unlucky enough to come into contact with a male polar bear, the odds are on that he'll try and make a meal of you. Female bears, particularly with cubs, are mostly just defensive – your chances of surviving an attack from them are better – but as a rule, polar bears take no prisoners.

The last hunt

In October 1995, Kyle Fumerton went on his first hunting expedition with his father. Being only six weeks old, Kyle didn't get out on the mountain with his dad, Shane, but stayed back at camp with his mum.

Shane Fumerton and his hunting companion, Bill Caspell, set out on mountain bikes early on a cold Monday morning. They planned to scale Mount Soderholm to hunt mountain goats.

When the two men failed to return from the hunt, Shane's wife, Carey, alerted rangers and local hunters. For five long cold days, Carey waited with baby Kyle at their mountain-base camp for news of her husband. Dangerous weather conditions hampered the search for the missing hunters. A helicopter sent into the area was nearly brought down by the high winds.

The search party found the bodies of the two hunters five days later, high on the mountainside, buried under 60 cm of fresh-fallen snow. A photograph recovered from a camera found at the scene of their deaths showed the two men with the carcass of a bull elk. They had set the camera's self-timer and caught the last few minutes of their lives on film.

> They caught the last few minutes of their lives on film.

As the hunters were de-boning the elk, they were charged by a grizzly. Bill Caspell was in the process of loading his gun when he was killed. Shane Fumerton didn't even get time to reach for his gun.

MAN-EATERS AND BLOOD-SUCKERS

His severely battered body was rolled 200 metres down the mountain.

Trackers found evidence that a number of grizzly bears were in the area at the time of the attack. A sow (mother) grizzly bear with two cubs, who was seen wandering between the bodies of the elk and the men, was assumed to be the cause of the hunter's deaths. She was shot by park rangers.

Shane and Bill were both experienced woodsmen and Bill was a former conservation officer but despite know-how and skill, nothing could save the two men from so sudden and ferocious an attack. If you go into the woods, sometimes even care and skill are not enough. The wilderness is home to wild things who defend their territory as we would our own.

4) Ferocious Ferals and Dangerous Domestics

In the right (or wrong) conditions, any carnivore can be driven to eat human flesh. Man-eaters come in all shapes and sizes, and sometimes in places where you'd least expect to find them.

PIGGING OUT

The idea of 'Babe' gorging himself on human flesh seems unbelievable, but it does happen, and not only with wild pigs. Pigs aren't fussy about where their food comes from – they even eat turds!

HAVING A BOARING TIME

Marlborough, Queensland, 1984
Late one September afternoon, nine-year-old Alf Collins and his dad set out to shoot a wild pig. Alf's dad had spotted a big black boar on his property the day before, feeding on the carcass of a dead cow, and he was keen to get rid of it before it caused trouble. Feral pigs are a major pest in many parts of the world, but particularly in northern Australia.

MAN-EATERS AND BLOOD-SUCKERS

DID YOU KNOW?
In 1954, an Australian farmer had a heart attack while tending his pigs. By the time his body was discovered, the pigs had gnawed away one of his hands.

While Alf's three little sisters waited in the truck, father and son set out in search of the pig. Unluckily, the pig found them before they found it. Without warning, the 130-kg boar came charging out of the scrub. Alf dived for cover as the boar careered into his father's legs, knocking him to the ground, slashing and slicing with its tusks and teeth. Alf's dad lunged at the boar, trying to turn its head away as blood spurted from his right leg. The boar inflicted a deep and crippling gash to his hand. Desperate and bleeding, he called to Alf for help.

slashing and slicing with its tusks and teeth

Ferocious Ferals & Dangerous Domestics

Alf armed himself with a fallen branch and went to the rescue. Punching, kicking, and shouting, he bashed the pig with the stick. Distracted and surprised by Alf's attack, it hesitated for a moment before continuing to gore the bleeding man. Alf set to again, using all his strength to beat the hairy black boar across the back. Finally, he served it one almighty kick in the rump. The pig gave up, and retreated into the bush.

Alf managed to drag his seriously injured father back to the truck and drive him home. Mr Collins was taken to Rockhampton Hospital for blood transfusions and to have dozens of wounds sewn up. Though one of his hands was permanently damaged, Alf had saved his life. Alf was presented with three awards for bravery by the Queen in 1986.

> **DID YOU KNOW?**
> Pigs first went bush in Australia not long after the early settlers brought them out from England. Feral boars descended from these early escapees developed long tusks and bristling manes (that's why they are called 'razorbacks'). They are the scourge of the outback. But it's not only in Australia that feral pigs are a problem. Worldwide, there are 23 sub-species of feral hogs. In the USA there are two million feral hogs in Texas alone!

DIABOLICAL DOGS

Maruki Dinsdale, a 59-year-old New Zealand farm owned a cute little mongrel dog called Puggy. Th

MAN-EATERS AND BLOOD-SUCKERS

not a real working dog, Puggy kept Maruki company, following him around the family property at Te Puke.

One evening in April 1997, as Puggy and Maruki were walking beside a creek, they crossed paths with a wandering pit bull terrier. Pit bulls are one of the most troublesome dog breeds, and are involved in a high proportion of dog fights and attacks on people. The pit bull set upon little Puggy, so Maruki stepped in, attempting to break up the fight.

Family members heard frantic barking from Puggy and the savage snarling of the pit bull terrier, and raced to Maruki's aid. Though they killed the vicious dog, which was still savaging Maruki's body, it was too late. Maruki was dead. He had injuries to over 90 per cent of his body, and one of his legs was almost severed.

Dangerous breeds

Some dog breeds are more likely to attack people than others. Obviously, bigger dogs can inflict more damage than small ones, but certain types of dogs are specially bred to be aggressive and also need to be watched. In Britain, pit bull terriers are banned because of the number of vicious attacks they have been involved in. The following breeds are most likely to be responsible for serious attacks because of their size and temperament:

Pit bull terrier	Malamute
Rottweiler	Husky
German shepherd	Rhodesian ridgeback
Doberman pinscher	Bull mastiff
Akita	Giant schnauzer

Ferocious Ferals & Dangerous Domestics

BEST FRIEND OR WORST ENEMY

In 1996, 4.7 million Americans were bitten by dogs. Of these, 2851 were postmen but the majority of the victims – 2.8 million – were children. You may think a dog bite hardly makes the dog a man-eater, but 20 of those people who were attacked were killed. Many of the other victims sustained appalling injuries. In Australia, around 30 000 people are attacked each year. Again, it's mostly kids that get mauled, and many of the injuries are to their faces.

> it's mostly kids that get mauled

Avoiding dog attacks
- Always check with the owner before approaching a dog.
- Be on guard if a dog shows any sign of aggression (bares its teeth, snarls, holds its tail high) or if it stares and its hair bristles.
- Never run past a dog that may want to defend its territory.
- Never be provocative or challenge a dog by staring at it.
- If threatened by a dog, back away slowly.
- Keep clear of dogs in packs, and never try to separate fighting dogs.
- Keep your cool – try not to scream, wave your arms or run away.
- Leave dogs alone when they are eating.
- Be careful around a dog with puppies.

MAN-EATERS AND BLOOD-SUCKERS

Dancing with Dingoes

When Constable Dale Searle's utility broke down in the Simpson Desert, he knew he was in for a long night. But the night turned into a nightmare when dingoes began to stalk him on the 50-km walk back to his base at Bedourie.

Constable Searle armed himself with a branch and beat the dingoes off as they rushed at him, snapping and jumping on him in an attempt to bring him down. He carefully rationed his supply of matches, lighting one at a time to keep the dingoes at bay. All through the long night, the dingoes circled him. In the morning, when he finally made it to the main road, they watched from a distance as he boarded a passing tour bus.

Dingoes actually have more in common with wolves than with domestic dogs, and despite many people's attempts to keep them as pets, like wolves they are essentially wild. Azaria Chamberlain, the 10-week-old baby that disappeared from her family's tent in the camping ground at Uluru in 1980, was undoubtedly taken by a dingo. Babies are particularly at risk in the company of dingoes as they are a convenient 'bite-size'. Like all predators, dingoes usually prefer to give humans a wide berth, but when the paths of dingoes and people cross, trouble often follows.

Rats on the Rampage

A teenage daughter's screams alerted her mother to the fact that something terrible had happened to the

Ferocious Ferals & Dangerous Domestics

new baby. The seven-week-old baby had been put to sleep in her pram the evening before. The mother raced to the pram to find her baby daughter dead. During the night, half of the small babe's face, left arm, hands and fingers had been eaten away by rats. This rat's tale is not from Europe in the middle ages but from Launceston, Tasmania, 1969.

When it comes to man-eating, rats pose the greatest threat to humans. There are more rats in the world than people and they are incredibly resilient. Throw a rat out the window of a five-storey building and the chances are he'll hit the ground running.

Rats will eat anything, and have no fear of nibbling on larger animals. The top front teeth of rats are powerful incisors, capable of biting big chunks of flesh from their victims. Rats are also ferocious fighters – most cats and even some dogs will avoid tangling with rats.

> **DID YOU KNOW?**
> The Norwegian brown rat, which is common throughout the world, can grow to 45 cm in length (including its tail).

Every year 1000 people – mostly kids – are bitten by rats in New York City alone. Rats are incredible breeders, easily producing seven litters of up to 12 little rats in a year. In Asia, rats eat enough grain every year to feed more than 250 million people. Despite this, in many parts of India, rats are considered sacred. In Rajasthan a temple dedicated to the rat god Ganesa is inhabited by thousands of rats who are fed daily with a special food called *laddu* made of grain, sugar and milk.

MAN-EATERS AND BLOOD-SUCKERS

If they can't eat you, rats can still kill you. Rat bites are prime transmitters of diseases and carry over 20 diseases that are dangerous to humans. In the fourteenth century, rats brought the bubonic plague to Europe which wiped out a quarter of the population of Europe (25 million people) in the space of a few years.

> **Rat bites ... carry over 20 diseases that are dangerous to humans**

> **DID YOU KNOW?**
> In 1980 rats were discovered on the island of Runit in the West Pacific. These rats were descended from underground rats that had survived 43 atomic bomb tests carried out on the island in the 1940s and 1950s. Although the island was considered unsafe for humans for at least the next 25 000 years, the rats were flourishing.

CATS MEOW

In 1995, a 69-year-old cat lover who lived in Leiden, Holland was eaten by his pet cats. The 15 cats were

Ferocious Ferals & Dangerous Domestics

locked in the house with him when he died unexpectedly. Because he was a loner, it was some time before his body was discovered. The starving cats had been trapped in the house, and they had eaten most of the old man before police discovered him.

TOO MUCH MONKEY BUSINESS

Jane Goodall is a world-famous zoologist who spent many years researching the lives of chimpanzees at the Gombe Stream Research Centre in Tanzania. Occasionally, she had to put her small son, Grub, in a cage to protect him from curious chimpanzees. When they have the opportunity, chimpanzees feed on the young of other primates – usually baboons and red colobus monkeys – but have been known to kidnap human babies. In one incident in Tanzania, a baby was rescued from chimps but not before its limbs had been partially eaten.

> chimpanzeees ... have been known to kidnap human babies

THE LAST WORD ON MAMMALS THAT MAUL

One of Africa's biggest man-*killers* (rather than *eaters*) is the hippopotamus. A hippo's jaws are so powerful it can bite a man in two. Hippos are territorial and if you enter one of their swimming holes to cool off, they won't be pleased to share it with you.

Bumping into an elephant when he's having a bad

MAN-EATERS AND BLOOD-SUCKERS

day can have unhappy results too. A tourist who stopped his car in an African wildlife park to offer an elephant a bun was crushed to death. The bun-hating elephant seized the man with its trunk and tossed him into the air before kneeling on him.

Any healthy, right-thinking animal will defend its territory if you charge uninvited into its home. Bulls, buffaloes and bisons are all capable of goring unwary intruders to death. In Yellowstone National Park more people are gored by bison than are attacked by grizzly bears.

Even if an animal isn't a meat-eater, it doesn't mean you're safe in its company.

5 Reptiles

HOME FOR THE HORRORDAYS

Twelve-year-old Peta-Lynn Mann loved coming home for the holidays. Her parents ran a safari business at Channel Point on the Timor Sea, 180 km south-west of Darwin, where the countryside is wild and ruggedly beautiful.

In April 1981, when Peta-Lynn came home from boarding school, Hilton Graham decided to treat her to a tour. Hilton was a business partner of her parents' and had always shown a friendly interest in her. He had even taught her to drive when she was only eight years old.

Hilton drove Peta-Lynn 20 km south to where a boat was kept for touring, and they spent a happy afternoon exploring the mangrove swamps. At nightfall, as they were heading back to the four-wheel-drive truck, their boat accidentally grounded on a sand bank. Hilton climbed out to push it clear. As he jumped into the shallow water, his pistol fell from its holster and he knelt down to search for it in the muddy water. That's when the crocodile took its opportunity.

Hilton barely had time to raise an arm in defence before a snapping bite broke it in two places. He struggled to his feet but the four-metre crocodile lunged again, this time closing its powerful jaws

MAN-EATERS AND BLOOD-SUCKERS

around his right thigh and dragging him into deeper water.

Peta-Lynn watched in horror, but fearlessly leapt into the water when Hilton cried for help. Grabbing his uninjured arm, she dug her heels into the mud and pulled with all her might. The crocodile – not to be so quickly outwitted – went into a death roll, sweeping Peta-Lynn off her feet and dragging both her and Hilton underwater.

They whirled and thrashed in the murky water, but Peta-Lynn wouldn't give in and she wouldn't let go. Regaining her footing, she managed to get Hilton's head above water and dragged both man and crocodile back to the bank. A few steps from the shore, the crocodile let Hilton go but as he staggered away, it lunged out of the water again, closing its huge jaws around his buttocks. For a moment, it looked as if the crocodile was going to win. But Peta-Lynn was stubborn. Wrenching Hilton from the mouth of the croc before it could establish its grip, she dragged him up the bank.

> Peta-Lynn wouldn't give in and she wouldn't let go

Peta-Lynn found a safe spot for Hilton 50 metres from the water, and raced back to the truck. With incredible presence of mind, she drove into the swamp, got the injured man into the truck, and drove back to the safari base camp. There was no one there, so Peta-Lynn covered his wounds in antiseptic powder and wrapped him in a sheet. Next, she radioed ahead, helped Hilton back into the truck, and set off for Darwin.

The following year the Queen awarded Peta-Lynn

A Bengal tiger (*panthera tigris tigris*) with remains of samba deer breakfast. The Bengal lives in tropical forests, bamboo thickets and reed beds in India, Nepal and northern SE Asia. It is an endangered species

Above: Crocodile hunting, Northern Territory, Australia

Above right: Leeches feeding on a bushwalker's foot

Above far right: Alf Collins and his father with a boar the same size as the one that Alf drove away (see Chapter 4); Alf was awarded three medals for the bravery he showed that day in saving his father's life

Right: Divers photograph a great white shark from inside a protective cage

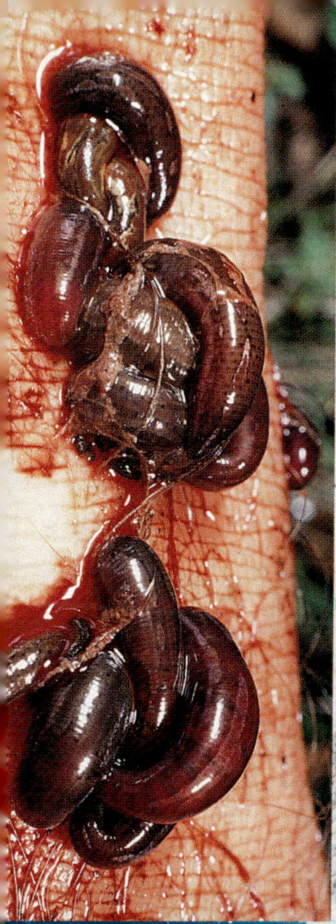
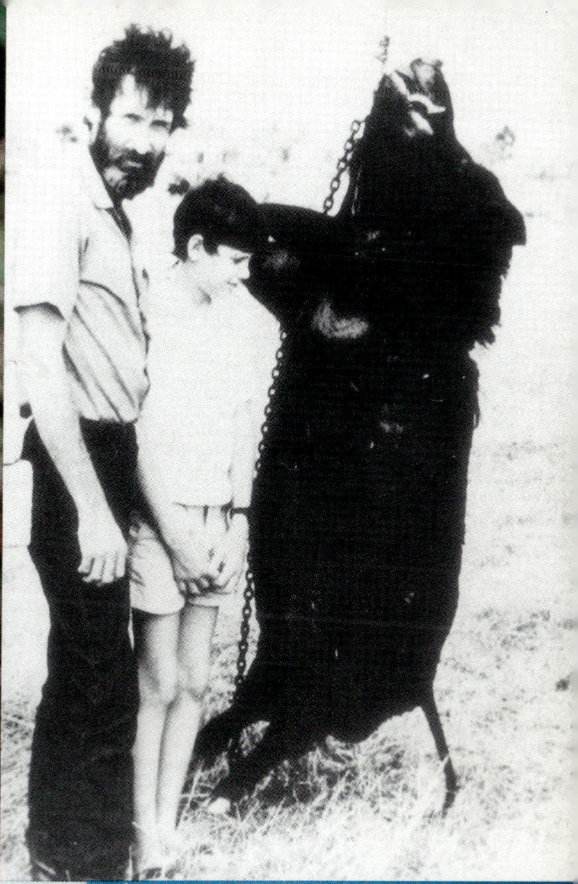
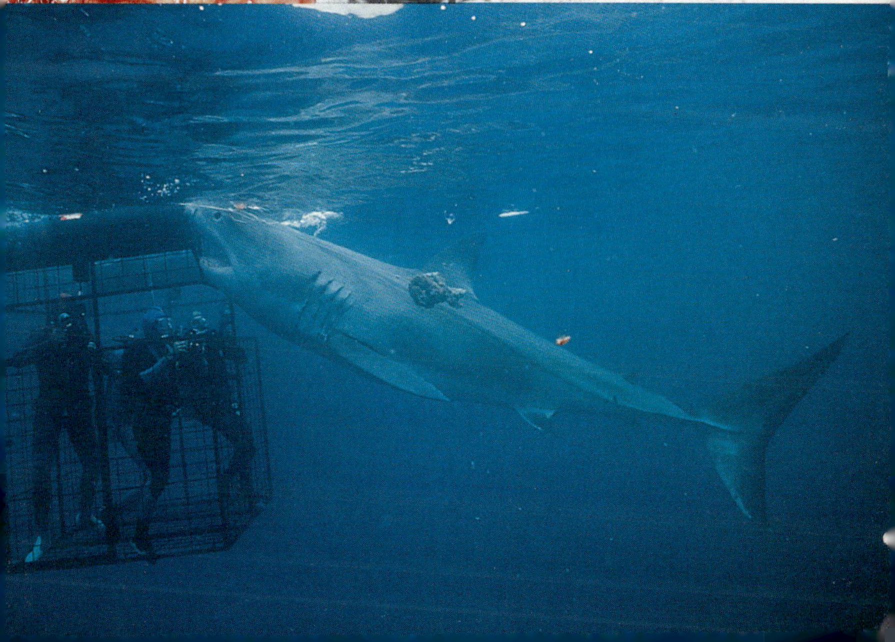

Grizzly bear

Reptiles

Mann the Royal Humane Society's gold medal for bravery and the Australian government's Star of Courage.

Though permanently scarred from the attack, Hilton recovered and went on to open a crocodile farm near Darwin.

RAMPAGE AGAINST REPTILES

There are 6000 species of reptiles in the world, but when you talk about *man-eating* reptiles, the list is short: saltwater crocodiles and their cousins, alligators and caimans; and the giant snakes – pythons, boa constrictors and anacondas.

There are two varieties of crocodile that inhabit the far reaches of northern Australia – the estuarine (saltwater) and the freshwater. Freshwater crocodiles are shy by nature, relatively harmless, and no more than three metres in length. When people talk of crocodile attacks, they mean saltwater crocodiles. But don't be fooled by their name – it's only used to distinguish them from their harmless cousins. Saltwater crocodiles also inhabit freshwater lakes, rivers and waterways. Fearless in their hunting methods, saltwater crocodiles can grow to over seven metres in length and more than a tonne in weight!

In Australia, cattlemen were the first to hunt crocodiles because the crocs preyed on their herds. After World War II, crocodile skins were in demand around the world, and the slaughter was on in earnest. The skins were used in the fashion industry to

> **The skins were used ... to make everything from shoes to suitcases**

MAN-EATERS AND BLOOD-SUCKERS

make everything from shoes to suitcases. Saltwater and freshwater crocodiles were massacred in their thousands to supply the skin trade, and by the late 1960s most of northern Australia's population of crocodiles had been wiped out. Between 1971 and 1974 legislation was introduced to protect the crocodile, and slowly their numbers began to recover. By 1990, 12 people had been fatally attacked.

> **DID YOU KNOW?**
> The ancient Egyptians worshipped the crocodile god Sobek. The city of Arsinoe was called Crocodilopolis by the Greeks because a special crocodile was kept in a lake there. The crocodile wore golden rings and bracelets inset with precious jewels on its legs. Every day the priests fed it meat, cakes and honey-wine. When a crocodile died, its body was mummified, just like a king's. Archaeologists have discovered thousands of crocodile mummies in sacred crocodile tombs.

A NILE CROCODILE'S EVENING SNACK

It had been a long hot afternoon and 15-year-old Jeremy Lloyd stripped down to his shorts for a quick swim in the Zambezi River. It was May 1987, and Jeremy was on holidays with his dad, Hugh.

Jeremy had been in the water for barely five minutes when a crocodile closed its jaws around him. He let out a gurgling scream, alerting everyone on shore of the attack.

Reptiles

Twenty-year-old Rupert Novis, a British army officer, jumped into the water and began wrestling with the crocodile. Jeremy's dad and Alexander Shaw, a holidaying student, quickly joined Rupert in the water, and they all battled the giant reptile in a desperate bid to save Jeremy's life. Alexander Shaw's arms were both broken by the lashing tail of the crocodile.

Despite the efforts of the three men, the crocodile had the boy firmly in its grasp, its teeth sinking into his thighs and buttocks. The situation became desperate as the crocodile moved off into deeper water. In a final courageous act, Hugh Lloyd forced his arm into the crocodile's mouth, giving Rupert

MAN-EATERS AND BLOOD-SUCKERS

Novis the opportunity to drag Jeremy away. The powerful jaws of the crocodile snapped shut and Hugh Lloyd's arm was severed at the elbow.

> **Keeping clear of danger in crocodile country**
> - Don't take risks. Avoid swimming in crocodile territory at dusk and at night, which is when they feed. Obey warning signs at all times.
> - Dogs are at risk too. Don't let them paddle or swim in the water.
> - Don't walk around at night without a torch.
> - Don't fish within three metres of the water's edge. When travelling in a boat, stay seated and don't trail your hand or any other part of you in the water.
> - Never interfere with a nesting mound.
> - If you come across a basking crocodile, head inland. Never come between a crocodile and the water.
> - Once a croc has you in its jaws, your chances are thin. Hanging on to some piece of the shore to prevent yourself being dragged into deep water for the death roll might give you a fighting chance. Jabbing something into its eyes or nostrils could distract it for a moment. A father in Queensland saved his 21-year-old daughter from the grip of a huge crocodile by gouging its eyes out.

CROCODILE ROCK 'N' ROLL

The family of crocodilia includes crocodiles, alligators, South American caimans and the smaller Indian gavial. All are tropical reptiles – they need warm climates to keep their cold blood flowing. They

Reptiles

are found in Africa, northern Australia, Asia and also in the Americas, both North and South.

Despite their reputation as man-eaters (they are the second biggest killer in Africa, next to hippos), crocodiles actually eat very little. Unlike mammals, these cold-blooded creatures get by on very little food. They only need a substantial feed once every two weeks, and eat about the same amount of food as a bird.

> They ... eat about the same amount of food as a bird

Like many of their dinosaur cousins, crocodiles are armour-plated – even bullets have trouble penetrating some parts of their skin.

Although they may look sluggish, crocodiles are deceptively swift. They can grab prey as big as a draught horse because of the incredible muscular strength of their jaws. No matter how hard you struggle, once a croc has a hold of you, your chances aren't good. If the first crushing bite doesn't kill you, it heads into deeper water and goes into a death-roll until it succeeds in drowning you. Crocodiles can stay submerged for up to an hour.

DID YOU KNOW?
An average crocodile will have around 6 kg of rocks in his stomach at any given time. Crocodiles swallow rocks and stones to aid their digestion. The stones also add extra weight to help keep them underwater.

MAN-EATERS AND BLOOD-SUCKERS

Sussing out serpents

The largest of all reptiles are the constrictors, a group which includes boas, anacondas and pythons. There is a lot of debate about just how big any of the giant serpents can grow. South American anacondas have been known to measure more than a metre around the middle and they can weigh over 220 kg. Anacondas are semi-aquatic – they are so huge they need the water to help hold them up – and moving on land is difficult for them. They feed on warm-blooded animals and are capable of eating a child or a small adult, but few people go wading alone through Amazonian swamps, so there is no proof of anyone being eaten by an anaconda.

People and pythons

The lives of pythons and people are deeply intertwined. Pythons have been slaughtered in huge

Reptiles

> **DID YOU KNOW?**
> There are many reports of anacondas reaching up to 20 m in length, but no one has yet been able to prove such a claim. Even half that length is pretty amazing. In fact, if you come across a snake that's more than 10 m long, there's a US$50 000 reward to be collected from the New York Zoological Society.

numbers to supply the fashion world with their beautiful skins. Their blood and gall bladder have been used in traditional medicines, and they have long been popular as exotic pets. Because of this, wild populations of pythons are listed as 'threatened' (one step down from being 'seriously endangered').

> **Pythons ... lunge with fantastic speed – so fast they can even catch birds in flight**

Snakes aren't really interested in eating people, but if you act or smell like their fortnightly feast, they may consider you fair game. Pythons are ambush predators; they lie in wait for their prey and then lunge with fantastic speed – so fast they can even catch birds in flight. Their teeth are needle-sharp to help in gripping their prey. Throwing their coils over their victim in a single movement, they squeeze tighter and tighter until all the breath is crushed out. When there is no longer any sign of life, the snake relaxes its grip and flickers its tongue along the body in search of the head – the end from which it starts to swallow. It takes days for the snake's body to digest its prey, and more than a week may pass before it sets out hunting again.

MAN-EATERS AND BLOOD-SUCKERS

WHEN A DREAM BECOMES A NIGHTMARE

It's fairly uncommon for anyone to wind up in a position of risk with a python, but accidents can happen.

Grant Williams and his brother Lamar loved reptiles. As little kids, the brothers caught grasshoppers, small lizards and turtles in parks and vacant lots near their home – a flat in the Bronx district of New York City. The two boys' great dream was to be herpetologists when they grew up. (Herpetologists are people who study reptiles and amphibians.)

The brothers bought their first snake when Grant was 13 and Lamar 10 years old. Over the next six years, they slowly built up quite a collection, including three pythons and over a dozen water, garter and milk snakes. They kept most of the snakes in tanks, but the biggest of the pythons, which they named 'Damien', roamed freely around the apartment. Their mother particularly disliked it and often begged the boys to sell it. Grant insisted the four-metre Burmese python was harmless, pointing out that it expressed no interest in even the family cat.

Mostly the boys fed the pythons live rabbits, but on one unlucky October day in 1996, 19-year-old Grant decided to give Damien a treat.

Giant pythons have amazing appetites and Damien was feeling very peckish when his master returned home with a live chicken under his arm. Pythons hunt by scent rather than sight so perhaps it was simply a case of mistaken identity – the python mistook Grant for his dinner.

Grant was discovered in a pool of blood by a neighbour who quickly summoned emergency workers. The medical officers pried Damien's 20-kg, 13-cm thick body from around Grant's neck and flung it into a nearby room. Grant was raced to hospital but was pronounced dead on arrival. Lamar, grief-stricken at the death of his brother, no longer dreams of becoming a herpetologist.

6 Creatures of Deep (and not so deep) Waters

THE BIG BITE

Rodney Fox was a keen spearfisher and one-time state champion. In 1963, he was competing in an event at Aldinga Bay, 55 km south of Adelaide. As he took aim at a morwong on the edge of Aldinga Reef, Rodney was charged by a white pointer shark.

Jaws clamped shut over his torso, tearing his wetsuit from shoulder to stomach. As he desperately reached backward in an attempt to gouge the shark's eyes, it released him. Unfortunately, his arm went into its open mouth, and razor-sharp teeth shredded his flesh.

As he thrashed his way to the surface, the shark followed. Desperate to keep away from its jaws, Rodney wrapped his arms and legs around his attacker. It plunged beneath the water, trying to shake him off. Breathless, Rodney swam to the surface.

The water was red with blood as the shark moved in for its final attack. Its jaws clamped around its target — this time, not Rodney but a marker buoy which was attached to his belt. With the buoy in its

Creatures of Deep (and not so deep) Waters

mouth, the shark plunged beneath the water again, dragging Rodney with it. Luckily the line snapped, and he floated to the surface in a state of shock, faint from blood loss.

The attack had been seen by the contest observers, and a boat was sent to rescue him. As he was taken onto the beach, his intestines spilled from his wetsuit. A bystander trained in first aid eased them back into place.

It took more than four hours of surgery and over 400 stitches to put Rodney Fox back together, but miraculously he survived. Rodney went on to study sharks and eventually worked as an adviser on underwater sequences for the film *Jaws*.

SHAPING UP TO SHARKS

White pointers are responsible for the highest number of attacks on man. The white pointer is also known as the great white, blue pointer or white death, and it is particularly dangerous because of its size. The longest white pointer on record was 9 m and weighed more than 4.5 tonnes. Even an average-sized one (half as big) could sever a person in two with a single bite.

Any shark over two metres is potentially dangerous to humans, but only 27 of 350 known shark species really present a threat to us.

Overfishing of sharks has seriously threatened

> **DID YOU KNOW?**
> Sharks have an endless supply of teeth. A tiger shark can produce and lose 24 000 teeth in its 10-year lifespan.

MAN-EATERS AND BLOOD-SUCKERS

their numbers. They are especially at risk because they are slow to reproduce. In South Africa, the great white shark is protected. Because so little is known about them, it is difficult to set limits on what level of shark fishing is appropriate.

There is still much about sharks that we don't understand. One thing we do know is that, as with land predators, humans are the dominant hunter. Each year we kill around 100 million sharks. Human deaths from shark attacks probably number fewer than 50. We're definitely coming out ahead.

> **Each year we kill around 100 million sharks**

Creatures of Deep (and not so deep) Waters

Keeping out of Jaws
- Avoid swimming alone and stay close to shore – sharks go for loners.
- Keep out of the water at twilight and at night time – sharks have the edge on you because of their keen eyesight, and they are more likely to be in feeding mode at night.
- Never enter the water if you're bleeding – sharks can scent blood in water and respond to a dilution of one part blood to a million parts seawater!
- Try not to look like a fish – shiny jewellery reflects light just like fish scales.
- Avoid wearing swimsuits with bright contrasting colour. Sharks are sensitive to contrast – uneven tanning can attract them too.
- Keep clear of areas where sewage or effluent is flushed into the sea or where anyone is fishing. Diving seabirds are a sign of heavy feeding activity, and sharks may be nearby. Areas between sandbars or where there are steep drop-offs are also hangouts for sharks.
- Avoid murky water.
- Don't swim with your pets.
- If you spot a shark, get out of the water as quickly and calmly as you can – too much splashing could attract its attention.

A HERO'S DEATH

Ray Boundy, Denis Murphy and Linda Horton were working on Broadhurst Reef, 90 km east of Townsville, when their trawler, *New Venture*, went down. They strapped a surfboard, lifebuoys and some

MAN-EATERS AND BLOOD-SUCKERS

foam padding together, hoping that a breeze would send their makeshift raft over to Lodestone Reef, where they could be picked up by other trawlers working the area.

The following evening they were nearly at Lodestone Reef when a four-metre tiger shark began to follow them. Curious, it nudged the floats and attempted to bite Ray's foot but he kicked it in the nose and it took off in fright. A few minutes later, a wave dumped all three people into the sea and the tiger shark moved in.

Murphy screamed as the shark bit his leg, severing it at the knee. With incredible courage, he turned to Denis.

'Well,' he said, 'it looks like that's it. You and Lindy bolt, 'cause he'll be back for the rest of me.'

Before they could stop him, Murphy swam straight at the shark. The water heaved as the shark closed its jaws around his torso and thrust his body out of the water. In agonising grief and fear, Linda and Ray paddled away in the opposite direction.

Around four a.m., the tiger shark cruised alongside Linda's lifebuoy. Suddenly, it seized the 21-year-old woman by the arm and body and shook her like a rag doll before disappearing into the sea with her. Ray was alone.

At daybreak, the shark returned yet again, circling Ray at a distance but coming closer with each pass. Ray was in sight of Lodestone Reef. Mustering the last of his strength, he surfed over the breakers on a piece of foam and was rescued after 36 gruelling hours in the water.

Creatures of Deep (and not so deep) Waters

SHARK TECHNIQUES AND OTHER TIDBITS

Worldwide, there are approximately 70–100 shark attacks every year. Between five and 15 people will die as a result of an attack. It's difficult to get really accurate figures because not all attacks are reported, but it's also important to keep fear of shark attack in perspective.

> You're 30 times more likely to be struck by lightning than eaten by a shark.

You're 30 times more likely to be struck by lightning than eaten by a shark.

The great hammerhead, shortfin mako, oceanic white-tip, Galapagos and some reef sharks have all caused their share of trouble with humans, but there are three especially dangerous species of sharks: the white shark, tiger shark and bull shark.

'Hit and run attacks' are the most common type of attack, and are usually a case of mistaken identity. Surfers who lose their legs to sharks are often victims of this sort of attack. The shark, perhaps mistaking the surfer for a seal, takes a single bite, realises its mistake and disappears. These sort of injuries rarely result in death, though if the bite is severe you may die of blood loss or drown.

DID YOU KNOW?
Fishermen in Fiji used to conduct a special ceremony to make the sea safe to fish in. Sharks were driven into a large net and 'shark-kissers' swam out, seized the nearest shark, turned it onto its back, and kissed its upturned belly. The fishermen believed that once a shark had been kissed, it would never attack a man.

MAN-EATERS AND BLOOD-SUCKERS

'Bump and bite' attacks and 'sneak' attacks are less common but more likely to leave you with fatal injuries. In bump and bite attacks, the shark begins by circling its prey, bumping the victim first. In sneak attacks, the shark circles for a while before making a direct assault. In both these types of attacks, the shark doesn't give up and will strike repeatedly. They usually happen in deep water to victims of sea disasters, such as sinking ships and plane accidents.

THE TATTOO MYSTERY

On 17 April 1935, Bert Hobson caught a 4.2-m tiger shark off Coogee Beach. It was such a whopper that he took it down to the Coogee Aquarium, where it was placed on display in a tank.

> **DID YOU KNOW?**
> Sharks may yet prove to be our greatest friends. Sharks have amazing immune systems — they suffer from few diseases and almost never develop cancer. Scientists are hoping to use sharks' blood to develop an anti-cancer drug.

A week later the shark seemed irritable. It had refused to eat for several days and on 25 April became agitated, bumping into the side of the pool. Something was definitely bothering the shark and, as it thrashed the water, it regurgitated a huge cloud of vomit. To the spectators' amazement, floating in the

Creatures of Deep (and not so deep) Waters

vomit was a human arm. The arm had a faded tattoo of two boxers on its forearm and a piece of rope tightly tied around its wrist.

Sharks are the pigs of the sea – they swallow whatever comes their way. A shark expert was called in, and he announced that the arm, which had spent eight days in the shark's belly, had been severed by something or someone other than a shark.

The arm was later identified as belonging to a local member of the underworld, James Smith. Police arrested two men, but no one was charged because nothing could be proved without a body.

Police believed James Smith was murdered. They suggested his body had been placed in a trunk but as his arm wouldn't fit, the murderers had hacked it off. The severed arm had been thrown into the sea where the tiger shark had swallowed it.

THE LONG ARM OF DEATH

In the late eighteenth century, two sailors cleaning the hull of a Dutch ship moored off the coast of West Africa were suddenly hauled into the sea by the long tentacle of a giant squid. A third sailor who was in the ship's rigging was also grabbed by the squid, but clung fast as his shipmates hacked the tentacle off. From the sample piece that was severed, the ship's captain estimated the whole arm to be 10–15 m in length. Giant squids are thought to have tentacles up to 28 m long but are very rarely sighted, as they live in the deepest parts of the ocean.

MAN-EATERS AND BLOOD-SUCKERS

PICKING ON PIRANHA

Famous horror stories describe shoals of piranha reducing busloads of tourists to skeletons in minutes.

Creatures of Deep (and not so deep) Waters

In reality, most piranhas are vegetarian. Their fearsome reputation is based on the fact that they have incredibly sharp teeth – so sharp that some South American Indians use them as razor blades!

> most piranhas are vegetarian

Red-bellied piranha, which are the most aggressive type, will occasionally attack an animal that is floundering in the water and trigger a feeding frenzy. A whole shoal of piranha will descend, each grabbing a bloody mouthful of flesh.

The South American candiru is much more disgusting in its eating habits. Tiny relatives of the catfish, candiru are as thin as a needle and less than 3 cm long. They feed on the mucus and blood of bigger fish but can also swim up a stream of urine if you're pissing into a river. They lodge themselves, by means of their sharp barbs, inside your penis or urethra. As they can make themselves at home in any opening of the body, swimming in waters infested with candiru is definitely a bad idea. You could be mistaken for a really big fish and find yourself host to these gruesome visitors.

THE PROBLEM WITH GETTING WET

No matter how much we love the water and feel at home in it, humans are land animals. There will always be some risks for us in water, of drowning if nothing else. Ironically, it's not the 'man-eaters' of the sea that we have to fear as much as our own ignorance and bad luck. In the seas surrounding Australia's coastline, the number of people that die from shark attacks is less than one per year, but an average of 92 people drown.

7 Parasites

THE MOST SUCCESSFUL MAN-EATERS OF ALL

Some creatures become dangerous to humans when their habitat is invaded, but for countless other creatures, the human body *is* their habitat. Thousands of invisible predators can share our bodies with us. Even in the most sterile and hygienic environments, we can't escape playing host to at least a few of them. It's not an idea that anyone welcomes. For many people, the fear of man-eaters is nothing compared to disgust at the thought of being dinner to a parasite.

GETTING AROUND ROUNDWORM

Veronica had a niggling feeling in the back of her throat before she sat down to dinner. She'd been feeling out of sorts all afternoon but was looking forward to eating the big plate of noodles her mother set before her. As the prickling sensation in her throat grew stronger, Veronica put her fork down and rubbed the side of her nose. Her brother, Alex, looked across the table at her in disgust.

'Veronica, you've got a noodle up your nose!' he cried, as a white, pencil-thin worm curled its way out of Veronica's left nostril. The family watched in horror as 20 cm of worm emerged from Veronica's

Parasites

nose and landed on the table beside her plate of steaming noodles.

Veronica's dinner guest was the intestinal roundworm, *Ascaris lumbricoides*. At any given time, one billion people around the world are playing host to the ascaris worm. If you're in good health and have a nutritious diet, you may not even notice it.

Ascaris worms usually stay put in your gut, but have been known to emerge from people's mouths, noses or even their belly-buttons! They can cause major problems as they rob their host of essential nutrients and may create a blockage or an infection in an internal organ.

Another type of roundworm is the whipworm. It makes itself at home in your colon (the lower part of your gut) and is shaped like a small whip. The whipworm thrusts its thin end into the colon wall and injects a fluid which converts the colon tissue into a sort of soup that it can then suck up.

People catch roundworms from food that is contaminated with worm eggs. If human excrement is used as a fertiliser, crops can be contaminated. When roundworm eggs hatch in your gut, the larvae pass into your blood system and find their way to your lungs, where they mature. Ten days later, you develop a cough. Coughing allows the worms to move from the lungs back to the gut, where they start laying eggs. Each female worm lays about 200 000 eggs every day, and up to 27 million in the course of her life. The eggs are carried out of your body in your faeces, and the cycle starts over again.

MAN-EATERS AND BLOOD-SUCKERS

A PICK AND MIX OF PARASITES

Unlike other predators, parasites have a vested interest in keeping their prey alive. Many parasites go their whole lives without being noticed. A parasite is probably keeping you company as you read this.

Most people's intestines have a quota of *Escherichia coli* (E. coli) bacteria making themselves at home. It's only when their numbers get out of control that we notice them.

Entamoeba gingivalis are tiny amoeba that live in your mouth and feed on white blood cells and leftover food that lies rotting in cracks and crevices. Ten per cent of healthy mouths and 95 per cent of unhealthy mouths play host to these microscopic bugs.

Tinea, or athlete's foot as it's commonly known, is actually a type of parasite too. It's a fungus that feeds on our dead cells, and lives in cosy damp folds of skin. Ringworm, which for a long time was thought to be a worm of some sort, is also a type of fungus that only shows up when it gets out of control. Most of the time, people and their fungi maintain a delicate balance.

> **DID YOU KNOW?**
> The word 'parasite' comes from the Greek word *parasitos*. In ancient Greece, a *parasitos* was a story-teller who received meals in exchange for telling amusing stories at dinner.

Parasites

THE TAPEWORM'S TALE

Eleven-year-old Dave Tully liked his steak rare and juicy. He thought nothing of helping himself to seconds of the tender, bloody sirloin that his dad cooked up. A neighbour who kept a hobby farm had brought around half a side of beef from a cow just slaughtered, and the whole family enjoyed a slap-up meal.

Over the next couple of weeks, Dave started to suffer from occasional stomach aches but it wasn't until two months later, when he turned to flush a turd down the toilet, that he decided something was really wrong with him. Dave took one look at what was floating in the toilet bowl and started screaming.

'My guts are falling out!' he howled, as his mother forced her way into the bathroom to see what was wrong. A 4-cm segment of thick, white, fleshy beef tapeworm was floating in the water.

DID YOU KNOW?
In Russia, a woman was found to have 90 metres of tapeworm in her intestines. Six individual tapeworms had infected her at the same time.

Dave's family, who were probably all infected, took a trip to the doctor for a dose of medicine that would clear their intestines of the unwelcome visitors.

Tapeworms are one of the oldest parasites – they've been with us for thousands of years and grow to incredible lengths. Beef tapeworms tend to be between five and seven metres long. One type of fish tapeworm that you catch by eating uncooked or

MAN-EATERS AND BLOOD-SUCKERS

pickled fish from the Baltic Sea grows to over nine metres.

Tapeworms are really effective feeders – they can absorb food through their entire body surface. Dave probably caught beef tapeworm from his dinner because the cow was not well cared for and the meat was undercooked. Raw pork, fish and beef can all contain the larvae of tapeworm. People can also catch a type of tapeworm from dogs.

Dave was lucky to discover he had tapeworm. Tapeworms are in you for the long haul – up to 30 years after you've eaten infected meat or fish, the worm may still be living in you. Using the hooks and suckers on their heads, they attach themselves to the wall of your intestine and grow in segments. Stories of tapeworms being tempted out of people by a plate of warm food or a bowl of milk are just a myth. People probably confuse wandering roundworms with tapeworms. Sometimes the only way to disengage a tapeworm is through surgery. If the head is not expelled, it may grow back.

> **Tapeworms are in you for the long haul – up to 30 years**

How to prevent parasites taking hold
- Wash your hands after going to the toilet, handling pets and before eating.
- Keep your fingernails short and clean.
- Wash fruits and vegetables before eating them.
- Keep your shoes on when playing outdoors.
- Steer clear of raw or undercooked beef, fish and pork.

Parasites

HOOKWORMS GET THEIR HOOKS IN

All other man-eaters combined make very little inroads on human life compared to parasites. One of the worst of these is hookworm.

Hookworms are blood-suckers that cause serious anaemia in their victims. Although only about 10 mm in length, each worm sucks up to 0.26 ml of blood from you each day. When you consider you could be playing host to 1000 worms, it adds up to a blood loss of half a litre every day! Vampires look really innocent in comparison.

> Vampires look really innocent in comparison.

The hookworm sinks its teeth into the wall of your stomach to make a hole through which it sucks your blood. As it sucks, it injects chemicals into you to stop the blood from clotting. Even when it lets go, the site it has bitten continues to bleed.

Hookworm larvae live in human excreta. The larvae attack the bare feet and ankles of people who come in contact with them and burrow up through the skin into the bloodstream where they are carried to the lungs. As with roundworm, when you cough, the worms come up from your lungs and travel down into your intestines, where they start their breeding cycle all over again.

Hookworms cause a lot of pain and suffering in countries where people don't have access to good sanitation. In the tropics, hookworm causes 50–60 000 deaths every year.

MAN-EATERS AND BLOOD-SUCKERS

WORMY WATERS

Amira checked the huge blister on her ankle before she went down to the waterhole. It was itchy and swollen. She had been feeling sick for a couple of days – dizzy and a bit feverish. A swim would be soothing – the idea of floating in the warm, golden water would take her mind off her sore ankle. But as she waded in, she felt a stinging sensation around the blister and it burst open, its milky contents washing away into the water.

When Amira went home, her mother checked the blister – which had quickly turned into a gaping ulcer – and found the head of a dead Guinea worm protruding from the centre of the wound. Its body was embedded in the muscles of Amira's legs and was probably 80 cm long, stretching from her foot to her hip.

> **Its body was embedded in the muscles of Amira's legs**

Amira had caught Guinea worm by drinking water infected with larvae and had unknowingly continued the cycle when she went swimming – as her blister burst open, millions of fresh Guinea worm larvae were

Parasites

released into the waterhole. The larvae had travelled through the wall of her gut and migrated into the muscles of her leg, where it grew and multiplied. When the Guinea worm finished its job of sending offspring into the world, it died. If left to rot and decay inside her body, it could cause serious infection and make her very ill.

Her mother gently teased the head of the Guinea worm out of the sore and wrapped it around a matchstick. Every day for the next few weeks, Amira's mother gently wound the body of the worm onto the stick, being careful not to break it. Because of their length (often well over a metre), it can take weeks to remove Guinea worms. Usually the worm makes its home in the legs of its victim, but if it decides to lodge in the torso or head it can be really dangerous.

Guinea worm isn't the only parasite that people can catch from contaminated water. Over 150 million people in the world have caught blood and liver flukes from bathing in or drinking infected water. Flukes are tiny flat leaf-shaped parasites about 10 mm long. Liver fluke are larger, reaching 30 mm in length. Both types of flukes feed on blood, are difficult to get rid of and cause swelling and pain in their hosts.

Apart from being disgusting, Guinea worms, flukes and intestinal worms cause a lot of unnecessary anguish. If everyone had access to clean water, guinea worms and flukes would become very rare and millions of lives would be saved.

8 Wings, Stings and Slimy Things

WINGED NIGHTMARES

On a hot summer night in 1991, 12-year-old Marisa dos Santos and her sister Marlene woke in fright. Unable to explain their unease, the girls went back to sleep. The morning light revealed both had been bitten a number of times by vampire bats.

Two months later, Marisa died in the Couto Maia Hospital in Salvador. The bats had not drained her of much blood, but they had infected her with rabies, an animal disease that is usually fatal for humans.

Marisa and her family lived in a small house in Apora, Brazil. In July 1991, intense logging activity in nearby forests resulted in hundreds of vampire bats being driven from their natural habitat where they fed on the blood of small forest mammals. As they flew in search of new homes, the bats quietly slipped in through the roofs of houses and fed on the local inhabitants.

Wings, Stings and Slimy Things

There are only three species of true vampire bats, all located in Central and South America. Most varieties of bats feed on fruit and insects.

Like Count Dracula, the vampire bat is incredibly stealthy. Once it has chosen a victim, it uses its heat sensors to find a piece of exposed flesh where blood vessels are close to the surface of the skin. In the case of a person, this is often an ear or toe. Unlike Dracula, who sucks blood through fangs, the vampire bat uses razor-sharp teeth to gently slice away a small area of skin. It does this so carefully you may not even notice, then laps up the seeping blood with its tongue. Vampire bats' saliva contains an anticoagulant which stops blood from clotting and keeps it flowing freely for a long time.

Often the bat is so gorged at the end of his feed that it is difficult for him to fly away until he has digested the blood. Although it generally takes only 30 ml from a single feeding session, in the course of a year a vampire bat can consume 23 litres of blood.

Bat colonies often choose to be near herds of cattle – horses are amongst their favourite victims. Incidents such as the Apora plague of vampire bats are rare, though once having discovered a tasty source of food, bats may return night after night for a quick drink.

THE FLYING DUSTMEN

Birds of prey are too smart to waste time preying on us. Stories about huge eagles kidnapping small children have never been proven and are probably

MAN-EATERS AND BLOOD-SUCKERS

myth; but one bird of prey that we do have an intimate and unusual relationship with is the vulture.

Vultures are intelligent birds that feed on carrion (dead animals). Their heads are bald which makes it easier for them to thrust into dead bodies. They also grow to incredible sizes – the Andean condor (a South American vulture) has a wing-span of over 3 metres! Despite the revulsion most people feel towards them, vultures are useful birds to have around. In many parts of the world they operate as flying dustmen, removing rotting flesh and disgusting refuse from urban areas – they even eat turds.

> vultures are useful birds

Like many skilled animal hunters that humans have feared and admired, most birds of prey are under threat of extinction.

Wings, Stings and Slimy Things

> **DID YOU KNOW?**
> In 1855, during the Crimean War between the British and the Turks, so many vultures turned up on the battlefields to eat the dead that special shooting squads were set up to fire at them so the wounded would not be eaten as well.

SKY BURIAL

The Zoroastrians of India lay their dead on towers which they call 'Towers of Silence'. Vultures feed on the bodies and reduce them to nothing but bones very quickly. The Zoroastrians believe the faster the human body disappears, the quicker the person is resurrected in the afterworld.

THE THING IN THE STING

There is one winged creature which really is life-threatening to humans. Every year, mosquitoes are indirectly responsible for the deaths of over one million children in Africa alone, through the spread of malaria.

Malaria is caused by a parasite that mosquitoes infect us with when they suck our blood. When the mosquito bites, parasites are injected into your bloodstream. As the parasites breed and die, their waste products are flushed into your body and cause a reaction that gives you a very high temperature – so high it may result in death.

Mosquitoes aren't born with malaria. They pick it up in their travels by biting a person, bird or animal

MAN-EATERS AND BLOOD-SUCKERS

that is already suffering from it. Mosquitoes can also infect humans with many other parasites and diseases, including Ross River fever, dengue, elephantitis and encephalitis.

Ticks, flies and fleas are just a few of the other bugs that enjoy snacking on our flesh and blood, although they don't like to live on us. Along with lice and mites, all these creatures are types of parasites. As with mosquitoes, what they leave us with causes more suffering than what they take away.

Despite the trouble they cause, all these creatures are an important part of the food chain, supplying fish, frogs, birds and other insects with a major part of their diet.

NOT-SO-NICE LICE

Mrs Barry had been cutting and curling hair all afternoon and was looking forward to shutting the shop. Alice was her last customer for the day.

'I must be tired,' she said, rubbing her eyes as she looked down on the top of the little girl's head.

Alice's hair seemed to be moving. As Mrs Barry lifted a lock of hair from the nape of the girl's neck, she noticed the skin was raw and red. As she blinked and looked closer, Mrs Barry realised Alice's hair was seething with tiny pale brown bugs.

Not many kids get through primary school without getting uncomfortably close to *Pediculus humanus capitis* – or head lice. Head lice aren't fussy – clean hair, dirty hair,

> **Head lice aren't fussy – clean hair, dirty hair, long or short, it all looks good to them.**

Wings, Stings and Slimy Things

long or short, it all looks good to them. They especially like friendly kids – the ones that put their heads together when they're chatting – as this makes it easier for them to move on. They have a claw on the end of each of their six legs to help them hang on once they've made the move.

An early sign of head lice is itchy scalp. This is not because of their wiggling bodies, but is caused by the saliva they ooze when they bite you. They pierce the skin of your scalp and spend seven minutes sucking their fill of blood. They especially like the tender skin at the nape of the neck and behind the ears, which is where you often first find their eggs. They lay the tiny golden eggs, called nits, close to the scalp to keep them warm. Each egg is firmly attached to a shaft of hair. Once they have hatched, the casing of the egg looks white or translucent.

> **DID YOU KNOW?**
> In some parts of South America, head lice are considered good luck because the lice immediately leave the body when their host dies. At least you know you're alive and kicking if you're scratching your head!

Before modern chemical treatments were invented, kids used to have their heads shaved when they became infested with head lice. It looks as if we may be heading back to the bad old days, as head lice seem to be developing a resistance to many chemicals.

TICKLES FOR TIM

Tim was in the kitchen helping his Mum get dinner ready when he first noticed how freckly his arms

MAN-EATERS AND BLOOD-SUCKERS

were. Since he'd turned 15, his arms had become really hairy but the freckles were something new. Tim wasn't the freckly type.

It was only when he got near the stove and started shifting pots and pans around, Tim noticed there was something special about his freckles. He picked up a lit match, and held it a centimetre above his arm. Tim's freckles were moving.

Tim was looking at *Phthirius pubis*, more commonly known as pubic lice or crabs. Unlike head lice, pubic lice prefer hair that is spaced further apart, especially where it's warm and moist. Their favourite home is the pubic area (between your legs), but armpits, eyebrows or any area of thick body hair will do. Pubic lice are usually caught through sexual contact, but you can also catch them from toilet seats or by sharing towels with people who are infected.

There are over 400 types of lice but they're a choosy bunch and only a few want to take up residence on us. Only head lice, body lice, pubic lice, itch mites (scabies) and face mites favour us over furrier friends.

> **DID YOU KNOW?**
> *Demodex folliculorum* (face mites) are tiny microscopic creatures that live on almost everyone's skin. Only 0.4 mm, these wormlike parasites can only be spotted by taking a skin scraping and putting it under a microscope. They rarely cause allergic reactions in their host and live in companionable silence alongside us.

Wings, Stings and Slimy Things

Leeches lick it up

Theo woke up during the night. Something was bothering him. Sleepily, he rubbed his eyes and ran his hand down the side of his neck. Screaming, he wrenched something slimy from his skin and flung it at the nearest wall, where it landed with a dull splat. When he switched the bedside lamp on, it illuminated a wall spattered with blood.

Theo had just rid himself of a blood-sucking leech. The week before, he had moved into a bungalow beside a marsh, little suspecting he was about to have a few unwelcome visitors.

There are 650 varieties of leech. Most can take on 10 times their own body weight in blood. When the head makes contact, it releases saliva which has a local anaesthetic in it. Some leeches have a muscular jaw filled with tiny teeth, others insert a proboscis that works like a hypodermic needle. As they suck, the body of the leech swells to five times its original size in about 20 minutes of furious bloodsucking.

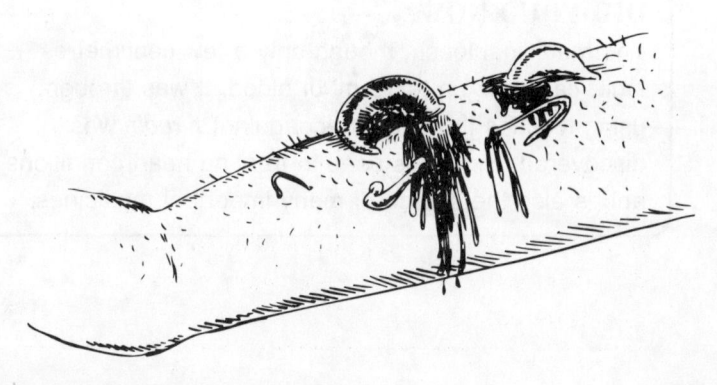

MAN-EATERS AND BLOOD-SUCKERS

Disgusting as leeches may seem, they are harmless and have sometimes proved to be of fantastic help to humans.

In the early eighteenth century, a French doctor, François Broussais, was one of the first people to use leeches in medicine.

> Leeches became a medical craze

He believed they could cure anything from coughs and colds to insanity. Leeches became a medical craze; doctors carried jars of them everywhere. Every year during the craze, medicinal leeches sucked 1.6 million litres of blood from the French people. In 1893 alone, 45 million leeches were gathered from ponds, lakes and marshes. By the turn of the century, leeches were nearly wiped out from many parts of Europe.

Leeches are still used in medicine today. They can help control the flow of blood during microsurgery and reduce swelling from bruising. Applying leeches to skin that is heavily congested with blood helps in the application of skin grafts.

DID YOU KNOW?
The medicinal leech, though only a few centimetres long, can suck up to 15 ml of blood. It was through using leeches that the anti-coagulant *hirudin* was discovered. Hirudin is used in treating heart conditions and is also the source of many important medicines.

9 Cannibals

THE MAKINGS OF A MURDEROUS MEAL

Richard could hear the other sailors murmuring but he felt too ill to listen to what they were saying. They had been adrift for 15 days with nothing to eat or drink. The night before he had been so desperately thirsty that he'd leant over the side of the dinghy and scooped up handfuls of sea water to drink. It hadn't done him any good – neither quenching his terrible thirst nor filling his hungry belly. He'd vomited it up and felt even worse.

'What's to be done? I believe the boy is dying,' hissed Captain Dudley. 'You have a wife and five children, and I have a wife and three children. Human flesh has been eaten before.'

Edwin Stephens, the first mate, couldn't look the desperate captain in the eye. He glanced across at the other seaman, Ned Brooks, and wondered if he had overheard the murderous suggestion.

'See what daylight brings forth,' he replied.

All through the night Richard lay groaning, deliriously ill and on the verge of a coma. The three other men tried to sleep but they were also in anguish from hunger and thirst.

In the morning, Captain Dudley scanned the horizon for any sign of a sail. Everywhere he looked was flat, monotonous, blue seawater. With ruthless

MAN-EATERS AND BLOOD-SUCKERS

desperation, he whispered an order to Edwin Stephens to hold young Richard's legs.

'Your time has come, my boy,' whispered Captain Dudley a moment before plunging his pen-knife into the jugular vein in the boy's neck. Blood spurted out. Captain Dudley caught it in the top of one of his navigating tools and thirstily drank it down.

THE LAST VOYAGE OF THE *MIGNONETTE*

The *Mignonette* was a small British yacht built in 1867 in Essex. Coincidentally, Richard Thomas Parker was born in a nearby village. The boy's life and the fate of the boat were to be gruesomely entwined.

On 5 May 1884 the *Mignonette* set out on its first and only ocean voyage. It was headed for Sydney, Australia with a small crew consisting of Captain Tom Dudley, first mate Edwin Stephens and two seamen, Ned Brooks and Richard Parker. Richard, only 17 years old, was an orphan. There was nothing in England to keep him and he was looking forward to a new life in Australia.

Unfortunately, the *Mignonette* was not designed for ocean travel. On 5 July she was caught in a gale in the south Atlantic, and sank. The four crew members found themselves cast adrift in a small dinghy, with the nearest land 4000 km away.

The nights were freezing and by day the men were scorched by the sun. Not only was there nothing to eat, they had no water. Eventually, they began to drink their own urine.

On 21 July, Captain Dudley took fate into his own

Cannibals

hands and murdered Richard Parker, convinced that the boy was going to die soon anyway. All three sailors drank the blood of the murdered boy and over the course of the following week, ate over half of his body. The conditions in the dinghy must have been disgusting with the half-eaten corpse rotting in the bottom.

On the morning of 29 July, as the men were 'eating their breakfast', the sail of a German sailing ship was spotted. After 24 days adrift, the men were rescued.

To the astonishment of the sailors, they were arrested and made to stand trial for the murder of Richard Parker. However hungry you might be, killing your shipmate to save yourself is still murder. Captain Dudley firmly maintained that if they hadn't eaten the boy, all of them would have died of thirst and starvation. Despite his plea, the three men were found guilty and sentenced to hang. But it was a difficult case – the jury, when it handed down its verdict, expressed confusion. It caused a sensation in the press and the Queen stepped in and pardoned the men on the condition they serve six months in prison.

Captain Dudley eventually made it to Australia, where he died of bubonic plague 15 years later.

HUNTING UP CANNIBALS

The case of the *Mignonette* is not the only example of people in desperate situations eating each other. Throughout history, stranded and shipwrecked sailors have gone to terrible lengths to survive. Escaped

MAN-EATERS AND BLOOD-SUCKERS

convicts, ignorant of how to gain food from the Australian bush, killed and ate each other in their desperation. Travellers stranded in freezing conditions and people suffering through famine have sometimes been driven to eat the bodies of the dead. But none of these situations are the images that we associate with cannibalism. Usually we think of cannibals as people who live in exotic places and pop stray missionaries in a pot for a feast.

In 1492 Christopher Columbus discovered the Caribbean and took tales back to Europe of a new world inhabited by bloodthirsty cannibals. Columbus had become friendly with the Arawaks and the Tainos people, and they assured him that the tribe next door, the Caribs, were a disgusting and repulsive bunch who ate human flesh.

Many anthropologists now believe that Columbus only heard what he wanted to hear. Sailors who discovered new continents often got the wrong idea about the people who lived there. Very few if any reports of cannibalism have been found to have any truth in them.

Cannibalism is the biggest taboo of all. Throughout history, when one group of people wanted to insult another they have accused them of cannibalism. As an insult, saying 'Your grandmother sucks eggs' is not a patch on saying 'Your grandmother eats babies'.

Accusing people of being cannibals or *anthropophagi* (the proper word for people who eat people) has also been used as a way of justifying taking away their right to be treated fairly. In 1926, gold miners in New Guinea were attacked by

Cannibals

Kaisenik tribesmen who resented the invasion of their lands by hundreds of prospectors. Because the miners claimed the tribesmen were cannibals and had eaten some of the prospectors, they received help from the administration and police to shoot scores of the Kaisenik people and blow up their village with dynamite.

Many cultures have used horrible punishments, especially for prisoners of war. Burial practices vary from one culture to the next and sometimes include eating parts of the dead. Some peoples believed that different parts of the body are sacred and full of magic and developed strange ways of expressing this, including scalping their enemies, shrinking heads and using them for decoration, or eating the eyes or liver of their enemies.

In 1773 Captain Cook visited the South Island of New Zealand. He wrote in his journal that officers had gone ashore and seen the head and bowels of a recently killed young man lying on the beach. The young man's heart was stuck on a forked stick and fixed to the head of one of their largest canoes. According to Cook, one of the Maori warriors brought the head of the youth on board and supposedly ate a small piece of its flesh. As with many stories of this kind, it is important to be a little bit sceptical –

DID YOU KNOW?

In 1970, a young American anthropologist working with the Lubgara tribe of Uganda was congratulated by the tribespeople for being something they thought was very rare – a white man that didn't eat African babies.

MAN-EATERS AND BLOOD-SUCKERS

Captain Cook may have only seen what he was expecting to see.

STAYING ALIVE

Robert Canessa scooped a handful of snow into a foil container and sat it on the roof of the plane alongside the strips of meat that were laid out to dry. He looked at the pieces of human flesh with a heavy heart. Despite his misgivings, he knew he would have to convince everyone to eat some. He had just heard on the tiny radio that the search for survivors of the missing Fairchild F227 aeroplane had been abandoned.

On 13 October 1972 the Uruguayan Air Force plane, carrying members of the Old Christians rugby team and their friends, had crashed into the side of a mountain in the Andes. The wing and tail sections were ripped off, and 10 passengers died immediately. Thirty people survived the crash.

The plane had been carrying only a small amount of food – chocolate, nougat, some tinned fish, wine and jam. Those team members who were able took care of their injured companions and struggled to make the

Cannibals

wreck of the plane warm and dry. All the young men showed incredible courage. They prayed together and cared for each other with compassion as they struggled against hunger, freezing weather, avalanches and the fear that they would never be rescued.

It wasn't until the tenth day that Robert Canessa cut strips from the body of one of the passengers who had been killed in the crash. He bent his head and prayed to God to forgive him before eating a small shred.

Some of the survivors couldn't bring themselves to partake of the only food that could give them life, and so they starved. Eventually everyone came to accept that the only hope of survival lay in eating the frozen bodies of the dead.

On 17 November, after a couple of failed attempts, Robert Canessa and two other boys set out for help. One of them returned to bring hopeful news to those left behind that there was still a chance of rescue. Thirty-four days later, Canessa and Nando Parvado reached civilisation. Helicopters were sent to rescue the survivors. Only 14 of the original 30 survivors were still alive on the mountainside. They had endured 71 days of frozen hell.

Glossary

anaemia: a condition where people have too few red blood cells and become very pale and tired

anthropologist: someone who studies the development of human culture and customs

anti-coagulant: a chemical that stops blood from clotting

black market: buying and selling of goods that are illegal

carnivore: any kind of animal that eats flesh

carrion: dead or rotting flesh

conservationist: someone who works at keeping animals, plants or their habitats safe and healthy, either as a hobby or as part of their job

endangered species: any species of animal that is in danger of being made extinct

environmentalist: similar to a conservationist but with an interest in the whole world

excreta: urine and faeces

feral: gone wild (used to describe an animal that was once kept by humans)

Glossary

food chain: all animals depend on other living things for food, so they are all connected like the links in a chain

habitat: the kind of environment in which an animal lives (e.g. forest, grassland, coast, etc)

larvae: the young of any insect that will metamorphise (change) as it becomes an adult

mammals: any warm-blooded animals that breast-feed their young

mangrove: a type of tree that grows in muddy swamps

omnivorous: something that eats all kinds of plant and animal foods

parasite: an animal or plant that lives in or on another species, and gets its nutrients from the host plant or animal

predator: an animal that hunts and preys on other animals

scavengers: animals that feed on leftovers, scraps and other dead animals

threatened species: an animal species that is at risk of becoming 'endangered'

Further Reading

The following books about man-eaters and bloodsuckers are full of ripping yarns and interesting info.

The Collins 'Eyewitness' series and the Golden Press 'Great Creatures of the World' series also include good books on animals.

The Beastly Book – 100 of the world's most dangerous creatures
by Jeanne K. Hanson
(Prentice-Hall, New York, 1993)
Full of fascinating bits and pieces about the scariest creatures you can come across.

Man-eaters of Kumaon
by Jim Corbett
(Oxford University Press, Oxford, 1993)
Unlike some big game hunters, Jim Corbett grew to love and respect the big cats, even setting up a reserve to protect tigers.

Never Cry Wolf
by Farley Mowat
(Bantam, New York, 1983)
This book changed people's ideas about wolves. It's a great read, but Farley Mowat admitted that although it is based on true experiences, it is really fiction. He left out quite a bit about the wolf's hunting habits because he was afraid of negative publicity.

Further Reading

Furtive Fauna – A Field Guide to the Creatures Who Live on You
by Roger M. Knutson
(Ten Speed Press, Berkeley, California. 1996)
From the amoeba to the fly, a very amusing look at the micro-critters that live on us.

Parasites
by A. J. Probert
(Kestrel Books, England, 1976)
A really gruesome look at intestinal worms.

Alive: the story of the Andes survivors
Piers Paul Read
(Secker & Warburg, London, 1974)
For older readers – a gripping account of the ordeal that drove the rugby team to cannibalism.

If you have access to a computer with **CD-ROM**, keep an eye out for the following.

CAT
Eyewitness Virtual Reality
(Dorling Kindersley Multi Media, London, 1995)

Sharks!
The Discovery Channel Multimedia
(Discovery Communications, 1994)

Bears and Pandas
Websters' Untamed Planet
(Webster Publishing and Teco Multimedia, Australia, 1995)

There are also many **Internet sites** worth a visit. I haven't listed them because they change or disappear so rapidly, but you could easily do a Web search to locate them.

Index

alligators 49
anacondas 49, 54
Apora, Brazil 76–7

Bardot, Brigitte 22
bats 76–7
bears 27–36
  black 31–2
  grizzly 27–31
  polar 33–4
boars 37–9

candiru 67
cannibals 85–91
cats, domestic 44–5
chimpanzees 45
cheetahs 10
Collins, Alf 37–9
Corbett, Jim 13, 95
crocodiles 47–53

dogs 39–41
dingoes 42

elephants 45–6

giant squid 65
Gévudaun 21
Goodall, Jane 45
Guinea worm 74–5

hippos 45
hookworm 73

Itard, Dr Jean 25–6

jaguars 15–17

leeches 83–4
lice 80–2
  head 80–1
  pubic 81–2
lions 12

malaria 79
Mann, Peta-Lynn 47–9
mosquitoes 79–80
Mowgli 25

Patterson, Colonel J.H. 12
pigs 37–9
piranha 66–7
pythons 49, 54–7

rats 42–4
reptiles 45–57
roundworms 68–9

Saint Ailbe 26
sharks 58–65
  tiger 62, 63, 64–5
  white pointer 58–9, 63
Sundarban, West Bengal 13

tapeworm 71–2
tigers 9–11, 13–15

vultures 77–9

Wolf Park, Haliburton 24
wolves 18–26
worms, intestinal 68–73